CAMBRIDGE
UNIVERSITY PRESS

University Printing House, Cambridge CB2 8BS, United Kingdom

One Liberty Plaza, 20th Floor, New York, NY 10006, USA

477 Williamstown Road, Port Melbourne, VIC 3207, Australia

314–321, 3rd Floor, Plot 3, Splendor Forum, Jasola District Centre,
New Delhi – 110025, India

103 Penang Road, #05–06/07, Visioncrest Commercial, Singapore 238467

Cambridge University Press is part of the University of Cambridge.

It furthers the University's mission by disseminating knowledge in the pursuit of
education, learning, and research at the highest international levels of excellence.

www.cambridge.org
Information on this title: www.cambridge.org/9781009168083
DOI: 10.1017/9781009168076

© Bnaya Gross and Shlomo Havlin 2022

First published 2022

A catalogue record for this publication is available from the British Library.

ISBN 978-1-009-16808-3 Paperback
ISSN 2516-5763 (online)
ISSN 2516-5755 (print)

Percolation in Spatial Networks

Spatial Network Models beyond Nearest-Neighbors Structures

Elements in the Structure and Dynamics of Complex Networks

DOI: 10.1017/9781009168076
First published online: June 2022

Bnaya Gross
Bar-Ilan University

Shlomo Havlin
Bar-Ilan University

Author for correspondence: Bnaya Gross, bnaya.gross@gmail.com

Abstract: Percolation theory is a well-studied process utilized by network theory to understand the resilience of networks under random or targeted attacks. Despite their importance, spatial networks have been less studied under the percolation process compared to the extensively studied nonspatial networks. In this Element, the authors will discuss the developments and challenges in the study of percolation in spatial networks ranging from the classical nearest neighbors lattice structures, through more generalized spatial structures such as networks with a distribution of edge lengths or community structure, and up to spatial networks of networks.

Keywords: spatial networks, interdependent networks, networks of networks, percolation theory, network resilience

ISBNs: 9781009168083 (PB), 9781009168076 (OC)
ISSNs: 2516-5763 (online), 2516-5755 (print)

Contents

1 Introduction

In recent decades network theory has been proven useful in describing many different complex systems from different disciplines, such as ecological systems (Paine, 1966; Pocock et al., 2012; Polis & Strong, 1996), epidemics (Pastor-Satorras et al., 2015; Wang et al., 2017), human microbiome (Gibson et al., 2016; Smillie et al., 2011), protein–protein interaction networks (Kovács et al., 2019; Li et al., 2017; Milo et al., 2002), finance (Stauffer & Sornette, 1999; Wei et al., 2014), climate (Donges et al., 2009; Fan et al., 2017; Ludescher et al., 2014), urban traffic (Hamedmoghadam et al., 2021; Li et al., 2015), and the human brain (Gallos et al., 2012; Moretti & Muñoz, 2013; Sporns, 2010). The resilience of complex networks is usually studied using percolation theory, which describes the robustness of the network under random failures or targeted attacks (Bunde & Havlin, 1991; Kirkpatrick, 1973; Stauffer & Aharony, 2018). Despite many advances in the field, especially in percolation of classical spatial structures such as lattices, most of the recent studies of more complex structures focus on nonspatial random networks, mainly because analytical solutions are easier to develop using different methods such as the generating functions formalism and mean-field approximations. However, when complex spatial networks are needed to describe real systems such as infrastructures (Latora & Marchiori, 2005; Yang et al., 2017), communication and transportation systems (Bell & Iida, 1997; Stork & Richards, 1992), or power grids (Yang et al., 2017), the preceding approaches fail and only limited analytical results for percolation theory are available.

In this Element, we will describe recent developments in the study of percolation in spatial networks. The Element is organized as follows: Section 2 will cover some basic principles of percolation theory for readers who are unfamiliar with this process. Section 3 will describe how the classical spatial structures of lattices were extended to more complex structures. We will also describe the structures of specific homogeneous and heterogeneous spatial networks and their behavior under the percolation process. In Section 4 we will describe the developments in recent years toward understanding the resilience of networks of networks and how they behave under percolation process in the presence of spatial constraints. Section 5 presents a new type of attack, localized attack on interdependent spatial networks, while Section 6 summarizes the conclusions.

2 Basic Remarks on Percolation
2.1 Classical Percolation

Percolation theory has been found useful to describe the structural phase transition of a network under random failures (Bunde & Havlin, 1991; Kirkpatrick,

1973; Stauffer & Aharony, 2018). In the percolation process, a fraction of $1 - p$ of nodes (site percolation) or edges (bond percolation) is randomly removed from the network, as shown in Fig. 1 for a 2D square lattice (of size $N = L \times L$). The functionality of the network is described by the relative size of the giant (largest) connected component (GCC), P_∞, and nodes that are not in the GCC are regarded as nonfunctional. Single networks experience a continuous second-order percolation phase transition where, above the critical percolation threshold, p_c, an infinite giant component (for an infinite

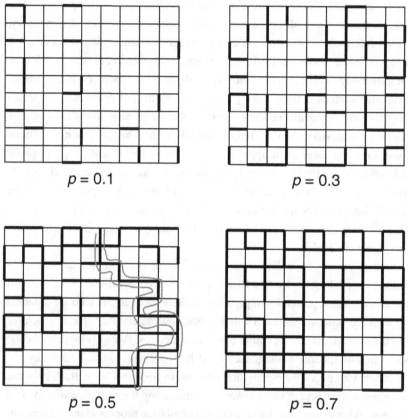

$p = 0.1$ $\qquad$ $p = 0.3$

$p = 0.5$ $\qquad$ $p = 0.7$

Figure 1 Demonstration of bond percolation on a 2D square lattice A fraction of $1 - p$ of edges is randomly removed from the network. At values lower than p_c ($p = 0.1$ and 0.3) only finite clusters exist and there is no path from one side of the network to the other (e.g., top to bottom). However, at $p = p_c = 1/2$ and higher ($p = 0.7$) a giant component that spans the entire network emerges, containing a path from one side of the network to the other. This phenomenon is known as a structural second-order phase transition, which usually characterizes percolation transition in a single network.
Source: Suki et al. 2011

network, $N \rightarrow \infty$) spanning the entire network exists, and the network is regarded as functional. Furthermore, in the case of spatial networks, the giant component contains a path from one side of the network to the other. In contrast, below the percolation threshold, the giant component is dismantled and only finite clusters remain (Fig. 1), and the network is regarded as nonfunctional. Except for the giant component, other important quantities can be evaluated during the process, such as the correlation length, ξ, which characterizes the typical length of finite clusters, and the susceptibility, χ, measured as the mean size of finite clusters.

Near (and at) the percolation threshold, a phase transition occurs and critical behavior is observed, represented by critical exponents:

$$P_\infty \sim (p - p_c)^\beta, \tag{2.1}$$

$$\xi \sim |p - p_c|^{-\nu}, \tag{2.2}$$

$$\chi \sim |p - p_c|^{-\gamma}. \tag{2.3}$$

In addition, several dimensional quantities exist that define different properties of the network. The first is the network dimension d relating the number of nodes in the network N to the linear size of the network L as $N = L^d$. In the case of random networks where L is not defined, the network is characterized by the small-world property $l \sim \log N$ (Milgram, 1967) (l is the chemical length, also called the minimal number of hops between two random sites), and the exponential increase of N with l indicates infinite dimension. Since the dimension of a random network is always larger than any network embedded in any d (N of random networks is larger with every L^d due to its exponential increase with l), it is impossible to embed random networks in space. Second is the fractal dimension $d_f < d$ of the giant component at the percolation threshold, $S = P_\infty \times N \sim L^{d_f}$. And finally, there is the dimension of the shortest path on the giant component d_{min}, which is defined as (Havlin & Nossal, 1984)

$$l \sim \langle r \rangle^{d_{min}}, \tag{2.4}$$

where l is the shortest path (called also chemical length, as mentioned), and $\langle r \rangle$ is the mean Euclidean distance between two sites at shortest path l. These dimensions together with the critical exponents satisfy a variety of scaling relations, such as $\beta + \gamma = \nu d_f$ and $d_f = d - \beta/\nu$ (Stanley, 1971), indicating that the critical exponents are not independent, and we can find all critical exponents by knowing only two of them.

The percolation phase transition and its critical exponents are significantly affected by the network structure. Interestingly, the concept of *universality class* groups together different systems with similar properties. For example,

Table 1 Some percolation-critical exponents. The full list can be found in Bunde and Havlin (1991).

d	2	6+
β	5/36	1
ν	4/3	1/2
γ	43/18	1
δ	91/5	2
d_f	91/48	4
d_{min}	1.13	2

the 2D square lattice and the 2D hexagonal lattice belong to the same universality class and share the same critical exponents (although with different p_c). Moreover, for all types of lattices an upper critical dimension $d_c = 6$ exists. Networks with dimension 6 or larger belong to the same university class, which is sometimes called the *mean-field* university class. Thus, while random networks do not follow the preceding definitions of d, d_f, and d_{min} since L and $\langle r \rangle$ are not defined, they take the values of a 6-dimensional lattice for all relevant scaling relations.

NAVIGATION IN NETWORKS

As a side remark, let us mention here the problem of *navigation* in networks, which shares some features of percolation. The problem of navigation in networks deals with the transmission of information between different nodes in the system (Milgram, 1967). In the basic model, at each step the information is transmitted to the closest node to the target. If the entire network structure is known at each step, the navigation route will simply be the shortest path, described by d_{min}. However, if only *local* properties are known, interesting routes characterized by different properties emerge (Kleinberg, 2000; Viswanathan et al., 1999). Moreover, additional interesting features appear when the networks are embedded in space (Hu et al., 2011; Huang et al., 2014), and the interested reader may study this topic as well.

Some of the most studied structures under the percolation process are random graphs. Random graphs with a Poisson degree distribution, $P(k) = \frac{z^k e^{-z}}{k!}$, where k is the degree (i.e., the number of links per node, with z being the average degree) are known as Erdős–Rényi (ER) networks (Erdős & Rényi, 1959; Erdős & Rényi, 1960). While space is irrelevant in random graphs, it was found that geometric metrics can provide important insight into the percolation transition (Boguñá et al., 2021). Percolation phase transition on ER networks is the same as for lattices of dimension equal or larger than 6 and also belongs to the mean-field universality class. The percolation properties of ER networks can be analytically solved using the generating functions formalism (Newman et al., 2001), and the size of the giant component can be obtained from the transcendental equation (Bollobás, 1985; Erdős & Rényi, 1959; Erdős & Rényi, 1960)

$$P_\infty = p(1 - e^{-zP_\infty}),\qquad(2.5)$$

with $p_c = 1/z$.

A similar well-studied random graph structure is the random regular (RR) network. In RR networks all the nodes have the same degree k_0 and the degree distribution is $P(k) = \delta_{k,k_0}$. Since the neighborhood of all nodes in ER and RR is the same and deviations from average are small, these networks fulfill translational symmetry and therefore belong to the same universality class and share the same critical exponents.

Another important random graph structure is the scale-free (SF) network. This structure describes many real networks such as the World Wide Web (Barabási & Albert, 1999), the Internet (Faloutsos et al., 1999), biological networks (Jeong et al., 2000), and airline networks (Colizza et al., 2006). Scale-free networks are characterized by a power-law degree distribution,

$$P(k) \sim k^{-\lambda}.\qquad(2.6)$$

In contrast to ER and RR networks, the existence of hubs in SF breaks the translational symmetry (in which every node sees a similar neighborhood), leading to a different universality class with different critical exponents. Scale-free networks with $\lambda \leq 3$ have been proven to be very robust with $p_c = 0$ due to the existence of hubs (high-degree nodes), which are rare but connect the network even after many random failures (Cohen et al., 2000).

While the generating functions approach describes random graphs well, it fails to describe spatial networks due to its local treelike structure assumption, which is usually not valid in spatial networks (Barthélemy, 2011; Gastner & Newman, 2006). Thus, analytical results for percolation in spatial networks are limited, with only a few results for the critical exponents in 2D (Den Nijs, 1979; Nienhuis, 1982) and a few analytical results such as $p_c = 1/2$ for bond percolation on a 2D square lattice (Fig. 1) (Sykes & Essam, 1964).

2.2 Network Resilience

While the classical percolation process is used to study network resilience under random failure, in recent decades percolation has been generalized for studying network resilience under other failure processes like targeted attack and localized attack. In a targeted attack, nodes are not randomly removed and instead central nodes or high-degree nodes are removed first. Albert et al. (2000) showed that in ER networks the percolation threshold is similar for random failures and targeted attacks. In contrast, scale-free networks are much more vulnerable to targeted attacks due to the existence of hubs. It was proven analytically by Cohen et al. (2000) that for random failures on SF networks with $\lambda \leq 3$, $p_c = 0$, while for targeted attack p_c is close to 1 (Cohen et al., 2001). That is, for random failures, unless one removes all nodes, the SF network does not collapse; however, for targeted attacks on high-degree nodes only a small fraction needs to be removed and the network collapses. One can take advantage of this property in order to efficiently immunize a population (or a computer network) by finding and immunizing the high-degree nodes (hubs) or super-spreaders in the population, which then block the spreading channels of the disease (Cohen et al., 2003; Liu et al., 2021). However, note that the case of targeted attacks on a spatial structure such as a 2D square lattice is trivial since the removal of a strip breaks the system into two clusters, which in the thermodynamic limit results in $p_c = 1$.

Localized Attack   Localized attacks in networks, in contrast to random failures, deal with the case where damage is not randomly spread in the network but rather localized in a given region of the network or a neighborhood of a given node. A simple scenario of localized attack in spatial structures is the creation of a hole of a given radius size r at the center of the network. Here also, the case of a spatial structure such as a 2D square lattice is trivial, and the critical radius size, r_c, where the system collapses is simply of the order of the system size, L, which in the thermodynamic limit $r_c \to \infty$ and corresponds to $p_c = 0$. Localized attacks have also been studied in random structures using the shell structure. Instead of creating a hole with geometric radius r, which has no meaning without space, a hole with shell radius l is removed. This means that a single node is removed, which corresponds to $l = 0$; then its neighbors are removed, which corresponds to $l = 1$, and so on until a fraction of $1 - p$ of nodes is removed. Shao et al. (2015) studied localized attacks on random graphs and found that the percolation threshold for random attack and localized attack on ER networks is the same. In contrast, SF networks are characterized by different behavior depending on the power-law exponent λ. For $\lambda < \lambda_c = 3.825$, the percolation threshold of localized attack is larger compared to that of random attack (i.e., localized attacks cause more damage compared to random failures), and for $\lambda > \lambda_c$ it is the opposite. This is since for small λ it is easy to find the hubs in the neighborhood of a random node, and localized attacks become more efficient in breaking the network compared to random attacks.

3 Beyond Classical Structures

The classical structure of spatial networks involves nearest neighbors connections with a fixed degree of each node, such as a 2D square lattice or a 2D hexagonal lattice or higher-dimensional lattices. These models have been extended to the case where the degree of nodes follows a distribution such as scale-free with $P(k) \sim k^{-\lambda}$ and the links were extended, if needed, to second and third nearest neighbors, and so on. In such cases the dimension of the shortest path (Eq. (2.4)) has been shown to be

$$d_{min} = (\lambda - 2)/(\lambda - 1 - 1/d). \tag{3.1}$$

Thus, for $d > 1$, due to the long-range links, the dimension $d_{min} < 1$. This result is the opposite of any lattice structure, which is known to have $d_{min} > 1$, and shows how even a small spatial change in the network structure can significantly change the properties of the network.

These studies (Ben-Avraham et al., 2003; Rozenfeld et al., 2002) broke free from the stiff shell of the classical nearest neighbors structures and opened a new direction for spatial embedded networks.

3.1 Homogeneous Structures

Studies of real-world spatial embedded networks show that, in many cases, the nearest neighbors linking structure, like lattices, is not realistic and that the Euclidean length of the networks' edges, r, follows a certain length distribution, $P(r)$. Bianconi et al. (2009) show that the distribution of edge lengths of the global airline network follows $P(r) \sim r^{-3}$. Lambiotte et al. (2008) show that the mobile phone communication network follows $P(r) \sim r^{-2}$ (see also Goldenberg and Levy (2009)), and Liben-Nowell et al. (2005) show that the spatial distribution of distances between social network friends follows $P(r) \sim r^{-1}$. To generally describe such systems, Daqing et al. (2011) developed a model of a random graph embedded in space where the nodes are the lattice sites with a Poisson degree distribution, $P(k) = \frac{z^k e^{-z}}{k!}$, and a power-law distribution of edge lengths,

$$P(r) \sim r^{-\delta}. \tag{3.2}$$

This model recovers the nearest neighbors structures of lattices for $\delta \to \infty$ and becomes a random graph for $\delta = 0$; see Fig. 2. Li et al. (2011) followed by Emmerich et al. (2013) studied this model and found that when the network is embedded in 2D, both its dimension and percolation properties change according to δ. For $\delta > 4$, there are very few long-range connections, and the network belongs to the universality class of percolation in a 2D lattice. For $2 < \delta < 4$, there are more long-range connections, and the percolation properties show new intermediate behavior different from mean-field, the dimension of the network increases continuously from $d = 2$ with decreasing δ, and the critical exponents depend on δ. Finally, for $\delta \leq 2$, there are many long-range connections, the dimension d becomes infinite, and the percolation transition belongs to percolation of random graphs. When the network was embedded in 1D, it was found that for $\delta \leq 1$, the dimension is infinite and the percolation transition is mean-field. For $1 < \delta < 2$, the dimension decreases and the critical

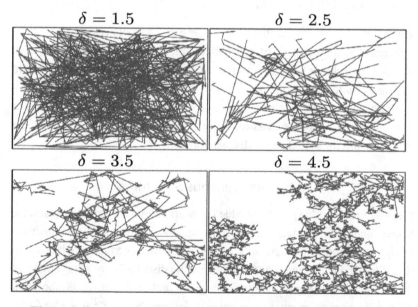

Figure 2 Demonstration of ER network embedded in space with power-law distribution of edge-length, $P(r) \sim r^{-\delta}$ The giant component of these networks embedded in a 2D square lattice at the percolation threshold with $\delta = 1.5, 2.5, 3.5,$ and 4.5 is shown. Since the edges are distributed randomly between nodes, the degree distribution is Poissonian, $P(k) = \frac{z^k e^{-z}}{k!}$ as in an ER network. As δ increases, the giant component has fewer long-range connections and becomes more affected by the constraints of the embedding space, as in 2D lattices with fractal dimension $d_f = 91/48$ (Bunde & Havlin, 1991; Stauffer & Aharony, 2018). **Source:** Li et al. 2011.

exponents depend on δ, and for $\delta > 2$, there is no percolation transition as in a 1D lattice. Schmeltzer et al. (2014) also studied this random graph embedded in space model, but with degree correlations. They showed that the percolation threshold can change in different ways according to the mechanism at which the degree correlation is induced in the model.

The ζ-Model   In contrast to our preceding examples of real-world networks that we have discussed so far, which are characterized by a power-law distribution of edge lengths, other real-world networks are characterized by an exponential distribution of edge lengths. Examples of such networks include transport systems or power grids (Danziger et al., 2016; Halu et al., 2014) and brain networks (Bullmore & Sporns, 2012; Ercsey-Ravasz et al., 2013; Horvát et al., 2016; Markov et al., 2014). The first model describing such systems is the Waxman model (Waxman, 1988). In the Waxman model, the distribution of edge length follows an exponential distribution,

$$P(r) \sim \exp\left(-r/\zeta\right). \tag{3.3}$$

Here, ζ is the characteristic length of the links, and the nodes are randomly distributed in space. Inspired by the Waxman model, a similar model with the

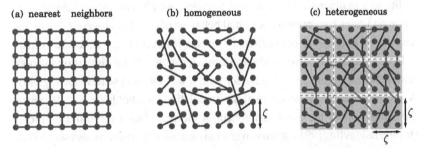

Figure 3 2D spatial embedded network models (a) The basic spatial models are the classical nearest neighbors models such as a 2D square lattice. **(b)** Spatial homogeneous models are extensions of the classical nearest neighbors models (in **(a)**), where the links are homogeneously distributed according to link-length distribution, $P(r)$. The example here demonstrates the ζ-model with the link-length distribution described in Eq. (3.3). **(c)** Demonstration of a spatial heterogeneous model which consists of communities of linear size ζ and distinct internal organization (brown edges). The intra-links can connect any pair of nodes within the communities. The communities are placed next to each other in a given lattice of dimension d, and only nearest-neighbor communities can be interconnected. The spatial effect is induced by the interconnections between neighboring communities in space (pink edges).

same edge-length distribution, Eq. (3.3), was developed (Danziger et al., 2016) where the nodes are not randomly distributed in space but rather organized as the sites in a 2D square lattice (see Fig. 3b) or other dimensional lattices. We call this model the ζ-*model* due to the spatial control parameter of the characteristic length ζ. Similar to the network having the distribution described in Eq. (3.2), the ζ-model having the distribution of Eq. (3.3) also recovers the nearest neighbors lattice structures in the limit of $\zeta \to 0$ and random graph structure for $\zeta \to \infty$. However, the main difference between both models, represented by Eq. (3.2) and Eq. (3.3), is the absence of long-range connections and having a characteristic length scale ζ in the ζ-model.

The ζ-model has been extensively studied in recent years under the percolation process (Bonamassa et al., 2019; Danziger et al., 2016, 2020; Gross et al., 2017). It was found that the percolation threshold of the two extreme limits $\zeta \to 0$ and $\zeta \to \infty$ is at the lattice threshold $p_c \simeq 0.5926$ and the random graph threshold $p_c = 1/z$ respectively, while the percolation threshold of intermediate values of ζ are between these limits (see Fig. 4a).

The varying length scale ζ in the ζ-model has been found to create a unique phenomenon of stretching which has not been observed before in the classical lattice model or in the power-law length distribution model (Eq. (3.2)). In short scales below ζ, each pair of nodes is likely to be connected with the same probability in a manner similar to that of a random graph with infinite dimension, while on scales longer than ζ a spatial behavior of $d = 2$ is observed due to the absence of long-range connections. This structural property leads the system to have a *crossover* between random behavior with infinite dimension in short scales (below ζ) and spatial behavior with $d = 2$ for long scales (see Fig. 4b). A similar crossover is observed at the percolation threshold; see Fig. 4c. Since the dimension of the giant component at the percolation threshold is fractal (Bunde & Havlin, 1991), a crossover in the fractal dimension is also seen. That is, at criticality, a mean-field fractal dimension $d_f^{MF} = 4$ is observed in short scales, and a 2D fractal dimension $d_f^{2D} = 91/48$ is observed in long scales (Fig. 4c).

This crossover phenomenon is also observed in the critical exponents close to the percolation threshold. An example of such a crossover is observed in the correlation length which is characterized by $\nu_{2D} = 4/3$ for 2D spatial structure and $\nu_{MF} = 1/2$ for random graphs and can now be described as (see Fig. 4d)

$$\xi \sim |p - p_c|^{-\nu} \quad \text{where} \quad \begin{cases} \nu = 4/3, & \text{for} \quad p^* < p < p_c \\ \nu = 1/2, & \text{for} \quad p < p^* \end{cases} \quad . \tag{3.4}$$

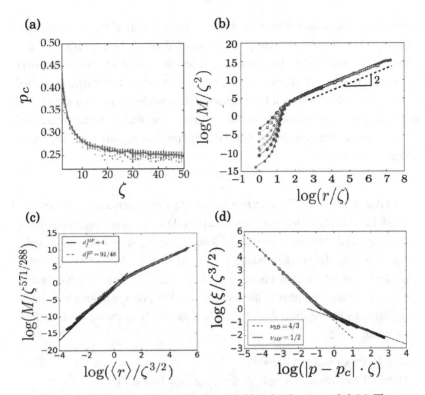

Figure 4 Crossover and critical stretching in the ζ-model (a) The percolation threshold p_c as a function of ζ. For $\zeta \to 0$ the structure is a 2D square lattice ($z = 4$) with $p_c \simeq 0.5926$, and for $\zeta \to \infty$ the structure is ER network with $p_c = 1/z = 1/4$. The percolation thresholds for intermediate values of ζ changes monotonically between these two values. **(b)** The dimension of the network follows the scaling $M = N \cdot P_\infty \sim r^d$. Away from criticality ($p = 1$) the dimension in short scales below ζ is infinite (the slope diverges) and for long scales above ζ the dimension is 2 (slope = 2). **(c)** At criticality, the dimension of the giant component is fractal with $d_f^{MF} = 4$ (mean-field fractal dimension) in short scales and $d_f^{2D} = 91/48$ (2D percolation fractal dimension) in long scales. Note that at criticality the mean-field behavior in short scales stretches up to $\zeta^{3/2}$. This is the *critical stretching phenomenon* where the mean-field behavior in short scales stretches to $\zeta^{3/2}$ instead of ζ at $p = 1$ shown in **(a)**. **(d)** Close to criticality the correlation length is long and characterized by $\nu_{2D} = 4/3$. Away from criticality the correlation length is short and characterized by $\nu_{MF} = 1/2$. The crossover between these two behaviors occurs at $\zeta^{3/2}$ due to the critical stretching phenomenon and the scaling is described by Eq. (3.4). **Source:** Bonamassa et al. 2019; Danziger et al. 2016.

Since ξ diverges at the percolation threshold close to p_c, large-scale behavior, very near p_c, is observed with $\nu_{2D} = 4/3$, and as p departs further ($p < p^* < p_c$)

from p_c, shorter-scale behavior is observed and the critical exponent changes to $\nu_{MF} = 1/2$ at the crossover point p^* which depends on ζ as $p^*(\zeta) = \zeta^{-1}$ (Bonamassa et al., 2019). Interestingly, the length scale at which the crossover occurs, ξ^*, depends nonlinearly on ζ at $\xi^* = \zeta^{3/2}$. This phenomenon is called *critical stretching*, and its interpretation is that, while away from the critical point ($p \gg p_c$), the random behavior is observed on scales shorter than ζ and $\xi^* = \zeta$, and at the percolation threshold the random regime stretches and is observed up to $\zeta^{3/2}$.

AN ALGORITHM FOR GENERATING A SPATIAL HOMOGENEOUS NETWORK

The following algorithm describes how to build a spatial homogeneous network on a d-dimensional lattice with a degree distribution $P(k) = \frac{z^k e^{-z}}{k!}$ and a general edge-length distribution $P(r)$. Start with the $N = L^d$ lattice sites (nodes) where the location of node i is $\vec{x}_i = (x_i^1, x_i^2, ..., x_i^d)$ and each coordinate is an integer number in the range $[0, L)$. The average degree is z, thus requiring us to distribute $E = Nz/2$ edges randomly in the network in the following way:

Step 1: Pick randomly a node i.
Step 2: Draw an edge length, R, from the distribution $P(r)$.
Step 3: Create a group $\tilde{R}$ containing every node j with Euclidean distance
$$d_{ij} = \sqrt{\sum_{u=1}^{d}(x_i^u - x_j^u)^2} = R.$$
Step 4: Connect node i to a randomly chosen node from group $\tilde{R}$.

Repeat steps 1–4 until E edges are distributed in the network.
Note: Since the value of R in Step 2 might not match an existing distance d_{ij} from the drawn node in Step 1, R should be taken to the closest value of an existing d_{ij}.

3.2 Heterogeneous Structures

In contrast to the homogeneous models discussed in Section 3.1, many spatial networks are heterogeneous, having a modular structure. For example, this structural property has been observed in transportation networks (Barthélemy, 2011; Guimera et al., 2005; Hajdu et al., 2019) and also in the brain (Barthélemy, 2011; Dosenbach et al., 2007; Gallos et al., 2012), and network stability is known to depend on the structure of the modules and the connections between them (Reis et al., 2014; Shai et al., 2015).

To model and better understand the resilience of such systems, Gross and Havlin (2020), Gross et al. (2020b), and Vaknin et al. (2020) developed a model of spatial modular networks. In this model, communities of linear size ζ are embedded in space next to each other as nodes in a 2D lattice, as can be seen in Fig. 3c. The internal organization of each community is like an ER network (i.e., every pair of nodes has the same probability to be linked), with average degree k_{intra}. The interconnections between communities are distributed among the community's four nearest neighbors communities with average inter-degree k_{inter} per node. This model is also motivated by the structure of infrastructures within and between cities in a country, where the cities are distributed in space with interconnections between them and a distinct internal organization within each city with easy connections between different parts of the city. Similarly to the ζ-model, this heterogeneous model is characterized by a random network behavior in short scales below ζ and a spatial behavior on long scales above ζ. However, it was found that in contrast to the homogeneous ζ-model, which experiences only a single percolation transition, the heterogeneous model experiences two percolation transitions (Gross et al., 2020b). The first (higher p_c) is the spatial transition when the communities become disconnected from each other at $p_c^{spatial}$, and the second transition is when each community breaks apart at p_c^{random}, as shown in Fig. 5a. Since each community is an ER network, the lower percolation threshold is simply $p_c^{random} = 1/k_{intra}$. The two transitions can

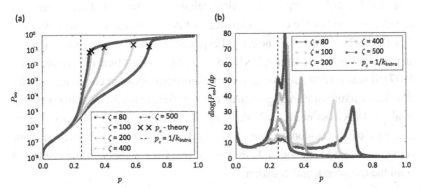

Figure 5 Two transitions in spatial modular networks (a) The giant component experiences two percolation transitions. The first percolation transition at the higher p_c is where the communities disconnect from each other (black **X**). This percolation threshold is obtained analytically from Eq. (3.5). The second percolation transition at the lower p_c is where each community breaks apart. Since each community is an ER network, it yields $p_c^{random} = 1/k_{intra}$. **(b)** The derivative of P_∞ with respect to p shows clearly the two peaks representing the two transitions in simulations. As ζ increases, the upper threshold gets lower, and in the limit $\zeta \to L$ the two thresholds merge.
Source: Gross et al. 2020b.

be clearly seen as peaks of the derivative of the giant component with respect to p (Fig. 5b). The upper transition gets lower as ζ increases, and in the limit of $\zeta \to L$ the two transitions merge.

Utilizing the analytical result of $p_c = 1/2$ for bond percolation in a 2D lattice, Gross et al. (2020b) were able to find the spatial percolation threshold analytically:

$$p_c^{spatial} = \frac{2\sqrt{1-2^{-1/Q}}}{1-\exp(-2k_{intra}\sqrt{1-2^{-1/Q}})}, \qquad (3.5)$$

where $Q = k_{inter}\zeta^2$ is the average number of interlinks emanating from each community. Moreover, since $p_c^{spatial} > p_c^{random}$, the size of the local giant component of each community, P_∞^{local}, at $p_c^{spatial}$ is not zero and can be found analytically as well:

$$P_\infty^{local}(p_c^{spatial}) = 2\sqrt{1-2^{-1/Q}}. \qquad (3.6)$$

Note that since $p_c^{spatial} > p_c^{random}$, at $p_c^{spatial}$ each community is above criticality, and therefore the critical stretching phenomenon found in the homogeneous model does not exist.

External Field Analogy   The critical behavior of the percolation transition is well described by the critical exponents of the percolation quantities such as Eqs. (2.1)–(2.3). However, the critical exponent δ, which describes the effect of an external field at criticality, has been studied as a ghost field (Reynolds et al., 1977). It was recently shown that interconnections between communities with random structures are analogous to an external field in magnetic systems (Dong et al., 2018; Gross et al., 2020a). Dong et al. (2018) studied the percolation transition of two random networks where a fraction r of nodes in each network is interconnected to the other. In the case of two ER networks, P_∞ can be found from the transcendental equation

$$e^{-zP_\infty}(r-1)+1-\frac{P_\infty}{p} = r\exp\left[\frac{Kp(e^{-zP_\infty}(r-1)+1-\frac{P_\infty}{p}-r)}{r}-zP_\infty\right], \qquad (3.7)$$

where z is the average degree of each network and K is the additional average degree of the interconnected nodes.

Dong et al. (2018) found that the interconnections remove the transition and the giant component is not zero at criticality ($P_\infty(p_c) > 0$). This behavior is analogous to the effect of an external field in magnetic systems and thus, the critical exponent δ can be defined for percolation, where r plays the role of the external field, as

$$P_\infty(r, p_c) \sim r^{1/\delta}. \tag{3.8}$$

The critical exponent δ was found to fit the value of the mean-field universality class $\delta_{MF} = 2$ and Widom's identity, $\delta - 1 = \beta/\gamma$, is satisfied.

Fan et al. (2018) studied the effect of interconnections between two lattices by adding interconnections to a fraction r of nodes from each lattice to the other. This structure can resemble a transportation network where the lattices describe the railroad networks of two countries, and the interconnections describe the airline routes between them (see Fig. 6a, b). Similar to the case of random networks, they found that for $r > 0$, at criticality the giant component is not zero (Fig. 6c), and the interconnections are analogous to an external field, as described in Eq. (3.8). Interestingly, the value of δ they found is in excellent agreement with the theoretical one of the 2D universality class, $\delta_{2D} = 91/5$, and Widom's identity is satisfied (Fig. 6d). Finding that the value of δ follows the

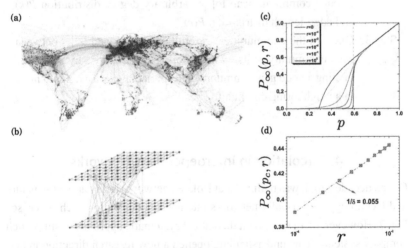

Figure 6 External field analogy in interconnected spatial networks (a) Transportation networks of railroads in countries and airline routes connecting them. **(b)** Modeling transportation networks with two lattices representing the railroad in each country, and a fraction r of nodes with random interconnections between the lattices representing the airline routes between countries. **(c)** The interconnections remove the transition, and $P_\infty(p_c) > 0$ for $r > 0$. Thus, the interconnections are analogous to the external field in magnetic systems. **(d)** The external field analogy allows us to measure the external field exponent, δ, defined in Eq. (3.8) that is found to have the value of the 2D universality class, $\delta_{2D} = 91/5$. **Source:** Fan et al. 2018.

universality class of the structure of the networks further supports the validity of Eq. (3.8) and demonstrates the principle of universality classes.

AN ALGORITHM FOR GENERATING A SPATIAL HETEROGENEOUS NETWORK
The following algorithm describes how to build a spatial heterogeneous network on a d-dimensional lattice. Start with the $N = L^d$ lattice nodes where the location of node i is $\vec{x}_i = (x_i^1, x_i^2, ..., x_i^d)$ and each coordinate is an integer number in the range $[0, L)$. The algorithm is described by the following steps:

Step 1: Divide the network into m communities, each community forms a d-dimensional cube containing ζ^d neighboring nodes, where ζ is the linear size of a community. The network now forms a d-dimensional cube of $(L/\zeta)^d$ communities (see Fig. 3c). Note that L/ζ should be an integer number.

Step 2: Build the internal organization of each community. The structure of each community can follow arbitrary degree distribution $P(k)$ and edge-length distribution $P(r)$.

Step 3: Once all the communities are built, randomly add links between each community to its $2d$ neighboring communities by randomly picking a node in a community and connecting it to a random node in a neighboring community; see Fig. 3c.

---

4 Percolation in Interdependent Networks

Over a decade ago, it was realized that isolated networks rarely appear in nature and technology, and usually networks interact and depend on each other; see Fig. 7. However, there was no mathematical systematic approach to study such complex systems. This understanding opened a new research direction in network science toward a network of networks and specifically *interdependent networks* (Buldyrev et al., 2010; Gao et al., 2011, 2012; Parshani et al., 2010). In interdependent networks, two types of links exist: ordinary connectivity links within the networks, as described in the sections 2-3, and *dependency links* between the networks. The dependency links between pairs of nodes in different networks imply that if a node at one end of a dependency link fails, then the node at the other end will fail as well, even if it is still connected to its own network. This process results in cascading failures and abrupt, first-order percolation phase transitions.

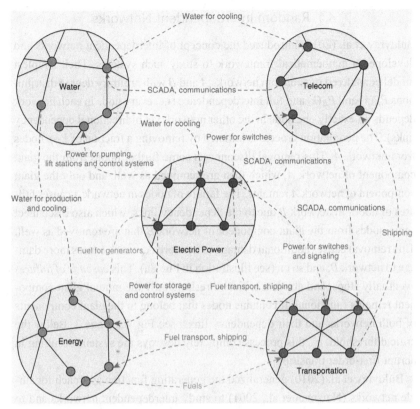

Figure 7 Interdependent networks The need of different resources for a system to function properly forms dependencies between different modern infrastructures, which causes cascading failures to spread between them.
Source: Gao et al. 2014.

Interdependent networks have been initially studied without spatial constraints. However, when spatial constraints have been considered, it has been found that the systems are more vulnerable to failures. Here we will discuss two models of how spatiality is introduced in models of interdependent networks. In the first model, the connectivity links are short range and fixed, as in a lattice, while the length of the dependency links changes and can be of a given characteristic length. In the second model, the dependency links are short-range ones that are fixed locally, and the range (length) of the connectivity links can vary. Both models show similar phenomena of extreme vulnerability, which are very different from nonspatial interdependent networks. However, before discussing them, let's briefly review the analytical results of random interdependent networks, which will help us to better understand the differences between spatial and nonspatial interdependent networks.

4.1 Random Interdependent Networks

Buldyrev et al. (2010) introduced the concept of interdependent networks and developed a mathematical framework to study such systems. Their simplest model considered two random networks A and B with arbitrary degree distributions $P_A(k)$ and $P_B(k)$ and full interdependence (i.e. every node in each network depends on exactly one node in the other network via bidirectional dependency links). The percolation process is initiated by removing a fraction of $1-p$ nodes from network A. This removal disconnects some further nodes from the giant component of network A, which then are removed as well, and only the giant component of network A remains. The failure of nodes in network A causes failures of nodes in network B due to the dependency links, which also disconnect some nodes from the giant component of network B that are removed as well. This removal causes additional damage to network A, which causes more damage to network B, and so on (see illustration in Fig. 8a). This *cascade of failures* eventually stops, and above a critical threshold only the mutual giant component remains functional. It contains nodes that belong to the giant components of both networks and their dependency links; see Fig. 8a, stage 3. Below the critical threshold, p_c, this process completely destroys the system, resulting in abrupt first-order transition.

Buldyrev et al. (2010) generalized the generating function approach for single networks (Newman et al., 2001) to study interdependent networks and to track the dynamics of the cascading failures and evaluate the final state of the system. To do so, let's first define some quantities: ψ' and ϕ' will be the fraction of active nodes in networks A and B respectively; the giant component of each network will then be $\psi = \psi' g_A(\psi')$ and $\phi = \phi' g_B(\phi')$, where g_A and g_B are the fractions of nodes belonging to the giant components of networks A and B if only ψ' and ϕ' nodes are active in each network respectively. The process is initiated by removal of a fraction of $1 - p$ of nodes from network A, which means $\psi'_1 = p$ and $\psi_1 = \psi'_1 g_A(\psi'_1)$. The damage now propagates to network B, and since every failed node in network A removes a single node in network B, we get $\phi'_1 = \psi_1 = \psi'_1 g_A(\psi'_1)$ and $\phi_1 = \phi'_1 g_B(\phi'_1)$. Following this approach, the giant component of each network can be evaluated at each step according to the following sequence:

$$
\begin{aligned}
\psi'_1 &= p & \psi_1 &= \psi'_1 g_A(\psi'_1), \\
\phi'_1 &= p g_A(\psi'_1) & \phi_1 &= \phi'_1 g_B(\phi'_1), \\
\psi'_2 &= p g_B(\phi'_1) & \psi_2 &= \psi'_2 g_A(\psi'_2)\dots \\
\psi'_n &= p g_B(\phi'_{n-1}) & \psi_n &= \psi'_n g_A(\psi'_n), \\
\phi'_n &= p g_A(\psi'_n) & \phi_n &= \phi'_n g_B(\phi'_n).
\end{aligned}
\tag{4.1}
$$

At the end of the cascading process, the system reaches a steady state, and no more nodes fail. This yields the conditions $\psi'_m = \psi'_{m+1}$ and $\phi'_m = \phi'_{m+1}$. If we denote $\psi'_m = y$ and $\phi''_m = x$, we get the following two coupled equations:

$$x = g_A(y)p$$
$$y = g_B(x)p. \tag{4.2}$$

The size of the mutual giant component μ_∞ that is composed of active nodes in both networks will be $\mu_\infty = xg_B(x) = yg_A(y)$, where $g_{A,B}$ can be obtained from

$$g_{A,B}(p) = 1 - G_{A0,B0}[1 - p(1 - f_{A,B}(p))]$$
$$f_{A,B}(p) = G_{A1,B1}[1 - p(1 - f_{A,B}(p))]. \tag{4.3}$$

Here, $G_0(x) = \sum_k P(k)x^k$ is the generating function of the degree distribution, and $G_1(x) = G'_0(x)/G'_0(1)$ is the generating function of the underlying branching processes.

For the case of two ER networks, the size of the mutual giant component can be described by the transcendental equation

$$\mu_\infty = p(1 - e^{-z\mu_\infty})^2 \tag{4.4}$$

and $p_c \simeq 2.4554/z$. In fact, Gao et al. (2011, 2012) showed that when n ER networks are fully interdependent in a treelike structure, the size of the mutual giant component is

$$\mu_\infty = p(1 - e^{-z\mu_\infty})^n. \tag{4.5}$$

This equation generalizes Eq. (2.5) of a single network, that is, $n = 1$. This equation has a trivial solution $\mu_\infty = 0$, where the networks are not functional. However, above p_c a new nonzero solution appears where the networks are functional. For $n = 1$, the new solution continuously emerges from 0, yielding a continuous second-order transition; while for $n > 1$, the new solution is not emerging from 0, yielding an abrupt first-order transition, as shown in Fig. 8b.

Partial Dependence   Parshani et al. (2010) studied the case of two random interdependent networks with *partial interdependence* (i.e. only a fraction of nodes in each network have dependency links). In their model a fraction of q_A nodes from network A depends on nodes in network B, and a fraction of q_B

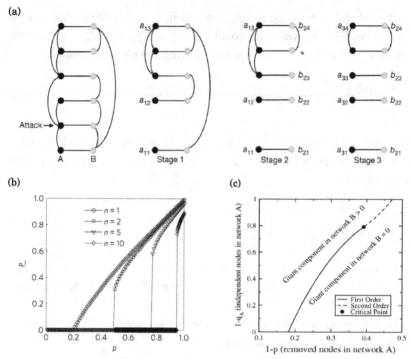

Figure 8 Cascading failures (a) Illustration of cascading failures in fully interdependent networks. A node in network A is initially removed by an attack. This removal disconnects nodes from the giant components of network A (stage 1). Next, the damage propagates to network B through the dependency links, and more nodes in network B become disconnected from its giant component (stage 2). Next, the damage propagates back to network A and so on until the cascading stops and only the mutual giant component remains (stage 3). **(b)** The percolation transition for n fully interdependent ($q = 1$) networks. The size of the giant component follows Eq. (4.5), with an abrupt transition for $n > 1$. **(c)** Percolation transition for two partially interdependent networks. For $q > q_c$ the transition is abrupt, while for $q < q_c$ the transition is continuous. **Source:** Buldyrev et al. 2010; Gao et al. 2012; Parshani et al. 2010

of network B nodes depends on nodes in network A. They showed that, in this case, the equations in (4.2) can be generalized to

$$x = p\{1 - q_A[1 - g_B(y)]\}$$
$$y = 1 - q_B[1 - g_A(x)p].$$
$$(4.6)$$

They found that for large values of q, $q > q_c$, the coupling is strong enough to yield an abrupt transition (first order); while for $q < q_c$, the coupling is too weak and the transition becomes continuous second order (see Fig. 8c).

4.2 Short-Range Spatial Connectivity and Varied Range Dependency

Interdependent r-Model   Wei et al. (2012) were the first to model and study percolation of spatial interdependent networks. Their model is composed of two networks, *A* and *B*, both with the same 2D square lattice structure. The length of connectivity links is one unit, that is, the bonds of the lattice; see Fig. 9a. The distance of the dependency links between the two networks is controlled by a distance parameter, *r*, allowing us to control the length of the dependency links. Each node is interdependent on another node in the other network (full interdependence) in the following way. A node *i* in a geometric position (x_i, y_i) in network *A* is interdependent with a random node *j* in a geometric position (x_j, y_j) in network *B*, and vice versa, if and only if $|x_i - x_j| \leq r$ and $|y_i - y_j| \leq r$, as illustrated in Fig. 9a. We will refer to this model as the *r-model* due to the spatial control parameter *r* of dependency links.

It was found that the distance, *r*, of the dependency links significantly affects the percolation transition, starting with the case *r* = 0, which is identical to percolation of a single network. This is since, for *r* = 0, any failure in one network leads to an identical failure in the other network, and no cascading

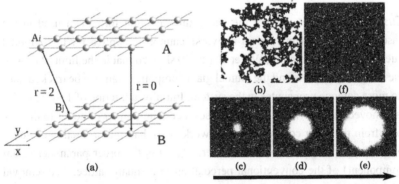

Figure 9 Spatial short-range connectivity with dependence of distance *r* – the *r*-model (a) The *r*-model considers two square lattices, *A* and *B*. Each node A_i in lattice *A* depends randomly on one and only one node B_j in lattice *B* via a dependency link (and vice versa), with the only constraint that $|x_i - x_j| \leq r$ and $|y_i - y_j| \leq r$. If node A_i fails, then node B_j fails, and vice versa. **(b)** For *r* = 0 the system is identical to a single lattice network having the same p_c and a continuous percolation transition. For low values of *r*, local failure cannot propagate, and the transition remains continuous as in a single network. **(c)–(e)** Once *r* exceeds the critical value $r_c \approx 8$, a spontaneous local hole can spread all over the system, similar to a nucleation process, and the system abruptly collapses. **(f)** As *r* increases further, the critical threshold decreases (see Fig. 12a) and the damage is randomly spread in the system instead of nucleating. **Source:** Wei et al. 2012

occurs. For low values of $r > 0$, the dependency links are not long enough for local damage to propagate, and the percolation transition remains continuous with a fractal structure of the giant component at the transition point (see Fig. 9b), similar to a single network. As r continues to increase, the damage can spread further, and the percolation threshold, p_c, increases approximately linearly with r, but the transition still remains continuous. However, once r reaches a critical value, $r_c \approx 8$, with the corresponding percolation threshold, $p_c \approx 0.738$ (see Fig. 12a), the dependency links are long enough, and a randomly appearing local hole propagates; and, governed by a nucleation process, the transition becomes abrupt, as shown in Figs. 9c–e. Once r increases above r_c, the spontaneous critical hole size required to destroy the system gets smaller and the percolation threshold decreases until saturating at $p_c \approx 0.683$ for $r \to \infty$; see Fig. 12a. Note that the preceding critical threshold values are for 2D square lattices and might change for other lattices. At the $r \to \infty$ limit, the critical point is characterized by a branching process, similar (but not identical) to the random case (Buldyrev et al., 2010), as shown in Fig. 9f. These results suggest that interdependent spatial networks are most vulnerable when the dependency length is in the intermediate range, thus highlighting the importance of understanding the spatiality induced by the dependency links.

Cascading Failures   While the generating function formalism used in the case of interdependent random networks cannot be applied for this r-model due to the spatial structure, Wei et al. (2012) showed that in the limit of $r \to \infty$ the cascading process and the mutual giant component can still be tracked until no more damage propagates in the system. Initially, a fraction of $1 - p$ of nodes is removed from network A. This causes a certain number of nodes to disconnect from the giant component of network A so that only a fraction of nodes $p_1 = P_\infty(p)$ remains functional, where $P_\infty(p)$ is the order parameter (giant component) of the conventional percolation in a square lattice. The removal of nodes in network A causes the removal of the dependent nodes in network B, which then causes failures of more nodes in network A, and so on. Similarly to Eq. (4.1) we can now track the evolution of the fraction of active nodes, p_i at step i, by

$$p_0 = p$$

$$p_1 = \frac{p}{p_0} P_\infty(p_0) = P_\infty(p)$$

$$\dots \tag{4.7}$$

$$p_i = \frac{p}{p_{i-1}} P_\infty(p_{i-1}).$$

When the cascading process ends, the recursive relation converges to a self-consistent equation, which describes the size of the mutual giant component,

$$x = \sqrt{pP_\infty(x)}, \qquad (4.8)$$

where $P_\infty(x)$ is the size of the giant component of a *single network* after removal of a fraction $1 - x$ of nodes, which is found via simulations. Once x is found, the size of the mutual giant component will be $P_\infty(x)$. For $x < x_c$ only the solution $x = 0$ exists, while for $x > x_c$ a nonzero solution appears, indicating a first-order transition; see Fig. 10. Thus, the percolation threshold satisfies

$$p_c = \frac{x_c^2}{P_\infty(x_c)}. \qquad (4.9)$$

Partial Dependence   Bashan et al. (2013) studied the r-model in the limit of $r \to \infty$ but with partial dependence. To their surprise, they found, using analytical arguments, that in contrast to random networks, which are characterized by a finite critical interdependence, $q_c > 0$, where for weak interdependence lower than q_c the transition becomes continuous and higher than q_c is abrupt,

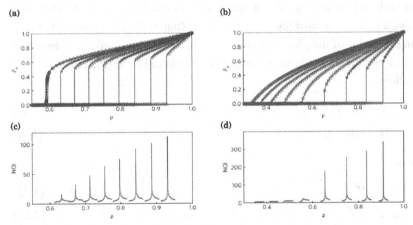

Figure 10 Spatial and nonspatial percolation transitions with partial dependence The plots are shown for $q = 0$ to $q = 0.8$ with steps of 0.1 (left to right). **(a)** Percolation transitions of the r-model (two lattices, Fig. 9a) in the limit $r \to \infty$ with different partial dependence. For any finite fraction of partial dependence, the transition becomes abrupt, yielding the critical coupling $q_c = 0$. **(b)** Percolation transitions of two partial interdependent random regular networks. In marked contrast to the interdependent lattices system, here $q_c \simeq 0.43$ and the system is much more resilient. The number of iterations (NOI) diverge at criticality for **(c)** $q > 0$ for lattices and for **(d)** $q > q_c$ for RR. This behavior is a typical characteristic of first-order transitions. **Source:** Bashan et al. 2013

in the case of two lattices, $q_c = 0$. That is, for any small and finite fraction of dependencies, the coupled r-model system will collapse abruptly, in contrast to random networks, as shown in Figs. 10a,b. Another measurement that supports these results is the number of iterations (NOI), defined as the number of steps within a cascading failure before it stops, where each step (iteration) includes transmission of damage from network A to network B and back. For first-order transitions, the NOI diverges at criticality, further supporting the result of $q_c = 0$ for lattices, which is in contrast to random networks (see Fig. 10c,d). This result highlights further the higher vulnerability of spatial networks compared to random networks. Later, Danziger et al. (2013) studied the r-model with partial interdependence, but for any value of r. They showed that while $r_c \approx 8$ for $q = 1$, when q decreases, the damage propagation for a given hole is reduced, and therefore larger holes are required to destroy the system, and r_c increases accordingly.

Plateau   The divergence of the NOI at criticality is characterized by a unique phenomenon called a *plateau*; see Fig. 11. Very close to criticality, the cascading failure slows down extremely and behaves as a branching process, where at each step on average only a single node in one layer causes a failure of a single node in the other layer (see Fig. 11c). Zhou et al. (2014) studied the scaling behavior of the plateau phenomenon in random interdependent networks close to criticality and found that the average plateau NOI, $\langle \tau \rangle$, and its std $\sigma(\tau)$ follow

$$\langle \tau \rangle \sim (p - p_c)^{-1/2}, \tag{4.10}$$
$$\sigma(\tau) \sim (p - p_c)^{-1} \tag{4.11}$$

and

$$\langle \tau \rangle \sim N^{1/3}, \tag{4.12}$$
$$\sigma(\tau) \sim N^{1/3} \tag{4.13}$$

at p_c.

The plateau phenomenon was also observed in the r-model, as shown by Wei et al. (2012), but a comprehensive scaling analysis of this phenomenon in spatial interdependent networks is still missing.

Healing   The r-model was further studied under the percolation process in the limit of $r \to \infty$ by Stippinger and Kertész (2014) but with a healing mechanism. The healing process works as follows: after removal of a node in a network

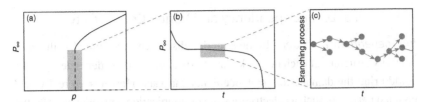

Figure 11 Demonstration of plateau phenomenon in cascading failure process (a) A first-order transition is characterized by an abrupt jump due to a cascading failure process. **(b)** At criticality the cascading failure slows down extremely, and a plateau phenomenon is observed. **(c)** The plateau phenomenon is caused by a branching process, where the cascading failure continues by a single node failure at each step. **Source:** Zhou et al. 2014

every possible pair of its neighboring nodes is getting connected independently with probability ω. This healing process makes the network more robust since finite clusters, which should have been removed, might connect again to the giant component due to the healing process. For $\omega = 0$ the regular interdependent percolation on lattices is recovered with abrupt transition at $p_c \approx 0.683$, as in Wei et al. (2012). However, as ω increases, p_c decreases accordingly, and at $\omega_c \approx 0.351$ the transition becomes continuous.

Networks of Networks Once the behavior of the r-model was understood for two coupled spatial networks, it was only a matter of time until it was generalized to more than two networks (i.e., networks of fully interdependent networks). Indeed, Shekhtman et al. (2014) studied the case of a treelike structure of n spatial interdependent lattice networks. They showed that while for $n = 2$, $r_c \approx 8$, when n increases, r_c decreases accordingly, and that for $n \geq 11$ r_c approaches 1, which means that for any spatial interdependence distance $r > 1$ the system undergoes an abrupt transition. Moreover, they applied the approach of Wei et al. (2012) and generalized Eqs. (4.8) and (4.9) for the case of n networks:

$$x = \sqrt[n]{pP_\infty(x)^{n-1}} \tag{4.14}$$

$$p_c = \frac{x_c^n}{P_\infty(x_c)^{(n-1)}}. \tag{4.15}$$

So far we have showed how spatiality affects the percolation transition when the distance is controlled through the length of the dependency links. In the next section, we will discuss an alternative model where the dependencies are in the same location ($r = 0$) and the distance is controlled through the length of the connectivity links.

4.3 Local Dependency and Varied Connectivity

Interdependent ζ-Model   In the r-model, discussed in Section 4.2, the connectivity structure of each network was short-ranged and the dependency links could bring the damage to a distance r that can vary. However, in real-world coupled networks such as electricity and communication networks, usually the dependency is in the same location, and only the connectivity within each network could be of longer range. To better describe the spatiality of such interdependent networks, Danziger et al. (2016), Vaknin et al. (2017), and Gross et al. (2021) studied the case of two fully interdependent networks where the dependency is local (i.e., in the same location); however, the structure of each network is generated as the ζ-model with characteristic link-length ζ, as described by Eq. (3.3). In this model, the damage can spread further via the connectivity links. This model is called the *interdependent ζ-model*, and it is a generalization of the single-layer ζ-model to the case of interdependent networks. In contrast, the dependency links are formed between pairs of nodes in the same geometric position in both networks; that is, node i in position (x_i, y_i) in network A depends on node j in position (x_j, y_j) of network B, and vice versa, if and only if $x_i = x_j$ and $y_i = y_j$, which is similar to the case of $r = 0$ of the r-model.

Danziger et al. (2016) studied the resilience of the interdependent ζ-model using percolation and found phenomena similar to those found in the r-model. For $\zeta < \zeta_c$ local damage cannot propagate, and the transition is continuous. However, once ζ is large enough (i.e., $\zeta > \zeta_c$), a random formation of a local hole can propagate throughout the system, by analogy to a nucleation process, and the transition is abrupt; see Fig. 12b, which is similar to Fig. 12a. For the case of $z = 4$, the interdependent ζ-model has the same amount of connectivity links as the r-model. For this case, $\zeta_c \approx 12$, which is larger than $r_c \approx 8$, indicating that the damage propagates more easily through the dependency links (r-model) than through the connectivity links (ζ-model). Also, the higher p_c of the r-model (Fig. 12a,b) indicates that dependency links propagate damage more easily. The source of this difference is probably the way both types of links transfer damage. Dependency links transfer damage *directly*, so if a node at one end of a dependency link is removed, the node at the other end will certainly be removed as well. In contrast, the connectivity links transfer the damage only in an *indirect* way, by disconnecting nodes from the giant component. Thus, removal of a node at one end of the connectivity link will not necessarily remove the node at the other end, since the node may still be connected to its giant component via another link.

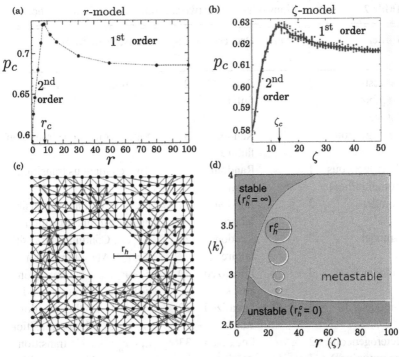

Figure 12 Percolation in interdependent spatial networks (a) Shown is p_c versus r for random failures in the r-model. For $r < r_c$ a randomly occurring local failure will not propagate, resulting in a continuous transition, while for $r > r_c$ a local failure will spread and abruptly destroy the system. **(b)** Shown is p_c versus ζ for random failure in the interdependent ζ-model. Similar to the r-model, here also two regimes exist. For $\zeta < \zeta_c$ a local failure will not propagate, resulting in a continuous transition, while for $\zeta > \zeta_c$ a local failure will spread and abruptly destroy the system. **(c)** Two interdependent ζ-model networks (red and blue). Both networks are interdependent, where the dependency links are formed between nodes in the same geometric position. Localized attack is initiated by removal of a circular hole of radius r_h at the center of the system. **(d)** Three phases exist in the $\langle k \rangle - r$ ($\langle k \rangle - \zeta$) phase diagram: there is an unstable phase, where the network spontaneously breaks and $r_h^c = 0$, and a stable phase, where $r_h^c = \infty$. Both of these phases appear in (a) and (b). Between these phases a novel metastable regime appears. In this regime, for $r_h < r_h^c$ the hole is too small and the failures will not propagate, while for $r_h > r_h^c$ the failure will propagate and the entire system will collapse. **Source:** Berezin et al. 2015; Danziger et al. 2016; Vaknin et al. 2017; Wei et al. 2012.

5 Localized Attack in Interdependent Networks

A new type of percolation phenomenon has been found for interdependent spatial networks, which is presented in both the r-model and the ζ-model. This phenomenon is called "localized attack," representing removal or failure of a node and its neighbors, and its next-nearest neighbors, and so on until a fraction

Table 2 Phase transitions of spatial network models under various processes

Model	# layers	Process	Page ref.	Phase transition
Nearest-neighbors (lattice)	1	Random failure	2	Continuous transition
Homogeneous	1	Random failure	8	Single continuous transition
Heterogeneous (communities)	1	Random failure	13	Two continuous transitions
r-model	2	Random failure	21	$r < r_c$: Continuous transition $r > r_c$: Abrupt transition
ζ-model	2	Random failure	28	$\zeta < \zeta_c$: Continuous transition $\zeta > \zeta_c$: Abrupt transition
r-model	2	Localized attack	29	$r_h < r_h^c$: No transition $r_h > r_h^c$: Abrupt transition
ζ-model	2	Localized attack	32	$r_h < r_h^c$: No transition $r_h > r_h^c$: Abrupt transition
Heterogeneous (communities)	2	Localized attack	33	$r_h < r_h^c$: No transition $r_h > r_h^c$: Abrupt transition

of $1 - p$ of nodes is removed. This scenario is very realistic for interdependent infrastructures, since in many cases, such as natural disasters or intentional attacks, failures are not randomly distributed, but a certain area is damaged, and this can be regarded and simulated as localized attacks on spatial networks.

Localized Attack in the r-Model Berezin et al. (2015) was the first to propose and study localized attacks as a percolation process and applied it on the r-model by removing a circular hole of radius r_h at the center of the system (see Fig. 12c). They found, surprisingly, that there exists a critical radius, r_h^c, below which the local damage of radius r_h does not propagate, while above it the local damage propagates through the entire system and the coupled system collapses. While the original average degree $\langle k \rangle$ of each layer is 4 due to the square lattice structure, the connectivity links can be diluted in order to reach a lower average degree or increased by also connecting the next-nearest neighbors. By studying the phase diagram in the plane $\langle k \rangle - r$ for the critical radius r_h^c, three regimes have been discovered, as shown in Fig. 12d. The first is an unstable regime for which the system collapses even for $r_h^c = 0$ (the same phase as found in Fig. 12a). The second is a stable regime for which $r_h^c = \infty$, which extends the left curve in Fig. 12a. Interestingly, between these two regimes a new metastable phase appears. In this phase a finite r_h^c exists, where circular damage with a radius larger than r_h^c will propagate in the entire system, causing system collapse, while localized attacks of circular damages with smaller radii will not propagate and the system will not collapse. Note that this phase corresponds to percolation with $p_c = 1$, since r_h^c does not depend on the system size. This vulnerability feature is unique and extremely dangerous, since finite damage can propagate and cause the collapse of the entire system. This is in contrast to percolation due to random failures or targeted attacks in interdependent networks, where the initial damage that is needed to cause system collapse scales linearly with the system size.

Localized Attack in the Interdependent ζ-Model   Vaknin et al. (2017) studied localized attack in the interdependent ζ-model. They initialized a localized attack of a circular failure of radius r_h at the center of the system (see Fig. 12c) and searched for the critical radius, r_h^c, at which the local damage propagates and the entire system collapses. Similar to the r-model they found three regimes, as shown in Fig. 12d. The first is an unstable regime for which the system collapses even for $r_h^c = 0$, the same value as that found in Fig. 12b. The second is a stable regime for which $r_h^c = \infty$, which includes continuation of the left curve in Fig. 12b. Finally, in between, a third phase appears, a metastable regime with finite r_h^c that depends on the model parameters. Note that Fig. 12d represents both interdependent models, r-model and ζ-model. This is because the processes are similar, but in the r-model the damage propagates through the characteristic length, r, of the dependency links, while in the ζ-model the damage propagates through the characteristic length, ζ, of the connectivity links.

Based on studies that showed that localized attacks in single ER networks have the same percolation threshold as random attacks (Shao et al., 2015; Yuan et al., 2015), Vaknin et al. (2017) found that for small values of ζ the size of the critical radius follows

$$\frac{r_h^c}{\zeta} \simeq \sqrt{z - z_c}, \tag{5.1}$$

where z is the average degree, and z_c is the critical average degree below which the network spontaneously breaks. They used the same arguments for the limit of $\zeta \to \infty$ and found the same scaling, suggesting again that

$$r_h^c \sim (z - z_c)^\beta, \tag{5.2}$$

with $\beta = 1/2$ describing the critical behavior of the process.

Localized Attack in the Heterogeneous Modular Model Localized attacks
have been also studied on an interdependent spatial heterogeneous (modular)
network model. Vaknin et al. (2020) studied the case of two interdependent
spatial heterogeneous network models, A and B. The heterogeneous structure
of each network follows the modular structure illustrated in Fig. 3c, where
each community is of linear size ζ; see Fig. 13a. Each node has a degree
k_{intra} of links within its community, a degree k_{inter} of links to other nearest-
neighbor communities, and the total degree is set to be $k_{total} = k_{inter} + k_{intra}$.
Another important parameter of the model is the interconnectivity ratio parame-
ter, $\alpha = \langle k_{inter} \rangle / \langle k_{total} \rangle$, which characterizes how significantly the communities
are interconnected to each other. Similarly to the interdependent ζ-model, the
dependency links are between nodes in the same geometric position in both
networks. Although it is usually very difficult to obtain analytical results for
percolation in spatial networks, for this case it was possible. Indeed, Vaknin
et al. (2020) utilized the generating function formalism in order to describe
the propagation of damage between and within the two networks and to find
the size of the mutual giant component during and at the end of the cascading
process. To describe the cascading process, we first need to describe a single
layer.

First, define $f_{i,j}$ as the probability that a randomly selected link, which passes
from a node in community i to a node in community j, does not lead to the giant
component. Following Newman (2003), we define the generating functions
$G_{i,j}(x)$ and $H_{i,j}(x)$ of the degree distribution and the excess degree distribution (i.e. the degree distribution of nodes at the end of randomly chosen link)
between community i and community j respectively. By defining p_j as the fraction of nodes that survived in community j as a result of an attack, $f_{i,j}$ should
satisfy

$$1 - f_{i,j} = p_j \left[1 - H_{j,i}(f_{j,i}) \prod_{l \neq i} G_{j,l}(f_{j,l}) \right], \tag{5.3}$$

where the index l is over the set of neighboring communities of community j,
including community j itself.

In a similar way, Vaknin et al. (2020) define g_i to be the fraction of nodes in
community i, which belongs to the giant component after removing a fraction
$1 - p_i$ of nodes in community i. Thus, g_i follows

$$g_i = p_i \left[1 - \prod_{j} G_{i,j}(f_{i,j}) \right], \tag{5.4}$$

where the index j goes over the set of neighboring communities of community i,
including community i itself. Note that Eqs. (5.3) and (5.4) are generalizations
of Eq. (4.3) for interdependent modular networks.

Using Eqs. (5.3) and (5.4), we can now track the cascading failures between the networks. To do so, we first denote the vector representation $\vec{f}$ with components $f_{i,j}$, the vector $\vec{p}$ with components p_i, and the vector $\vec{g}$ with components g_i. Now, Eq. (5.3) can be written as

$$\vec{f} = \vec{\Phi}(\vec{f}, \vec{p}), \tag{5.5}$$

and Eq. (5.4) can be written as

$$\vec{g} = \vec{\Psi}(\vec{f}, \vec{p}). \tag{5.6}$$

The initial attack is $\vec{p}(0)$, and we can now track the cascading failures in a similar way as described in Eq. (4.1) by following the time evolution of $\vec{f}(t)$, $\vec{g}(t)$, and $\vec{p}(t)$:

$$
\begin{aligned}
\vec{f}_A(2t) &= \vec{\Phi}_A[\vec{f}_A(2t), \vec{p}(2t)], \\
\vec{g}_A(2t) &= \vec{\Psi}_A[\vec{f}_A(2t), \vec{p}(2t)], \\
\vec{p}(2t + 1) &= \vec{\Psi}_A[\vec{f}_A(2t), \vec{p}(0)], \\
\vec{f}_B(2t + 1) &= \vec{\Phi}_B[\vec{f}_B(2t + 1), \vec{p}(2t + 1)], \\
\vec{g}_B(2t + 1) &= \vec{\Psi}_B[\vec{f}_B(2t + 1), \vec{p}(2t + 1)], \\
\vec{p}(2t + 2) &= \vec{\Psi}_B[\vec{f}_B(2t + 1), \vec{p}(0)], \dots.
\end{aligned}
\tag{5.7}
$$

At $t \rightarrow \infty$ the vectors $\vec{g}_A(t)$ and $\vec{g}_B(t)$ will converge to the mutual giant component μ_∞.

The size of the mutual giant component at the end of the cascade can be obtained by the same approach as that described in Eqs. (5.3) and (5.4). We now define $f_{i,j}^A$ as the probability that a randomly selected link, which passes from a node in community i in network A to a node in community j in network A, does not lead to the *mutual* giant component. Similar to Eq. (5.3), $f_{i,j}^A$ satisfies

$$1 - f_{i,j}^A = p_j \left[1 - H_{j,i}^A(f_{j,i}^A) \prod_{l \neq i} G_{j,l}^A(f_{j,l}^A) \right] \times \left[1 - \prod_l G_{j,l}^B(f_{j,l}^B) \right]. \tag{5.8}$$

$f_{i,j}^B$ satisfies the same equation but with exchanging indexes A and B.

The mutual giant component of community i, $\mu_{\infty,i}$, is the probability that a randomly selected node in community i survived the initial attack and has in both networks at least one neighbor that belongs to the mutual giant component. Thus,

$$\mu_{\infty,i} = p_i \left[1 - \prod_j G_{i,j}^A(f_{i,j}^A) \right] \times \left[1 - \prod_j G_{i,j}^B(f_{i,j}^B) \right]. \tag{5.9}$$

In the case of ER networks, $G_{i,j}(x) = H_{i,j}(x) = \exp\left[\langle k_{i,j}\rangle(x - 1)\right]$ and Eq. (5.4) takes the form

$$g_i = p_i \left[1 - \exp\left(-\sum_j \langle k_{i,j}\rangle g_j\right)\right],\tag{5.10}$$

and a self-consistent equation for the mutual giant component can be obtained:

$$\mu_{\infty,i} = p_i \left[1 - \exp\left(-\sum_j \langle k_{i,j}\rangle_A \mu_{\infty,j}\right)\right] \times \left[1 - \exp\left(-\sum_j \langle k_{i,j}\rangle_B \mu_{\infty,j}\right)\right].\tag{5.11}$$

Using Eq. (5.11), we can also evaluate analytically the size of the critical radius r_h^c by translating r_h to p_i by counting the fraction of lattice sites outside the hole of radius r_h in the damaged communities. Fig. 13b shows the size of the critical radius r_h^c as a function of α for different values of $\langle k_{total}\rangle$, and two regimes are observed. The critical values α_c were found when for $\alpha > \alpha_c$ we had a metastable regime, where a finite-size localized attack larger than r_h^c causes cascading failures, leading to system collapse. In this regime the critical radius r_h^c depends weakly on the interconnectivity parameter α. It was also found that in this regime, r_h^c is independent of the number of communities in the system (i.e., independent on the system size), meaning that $p_c = 1$. In contrast, for $\alpha < \alpha_c$, the critical attack is $r_h^c \sim 0.5L$ (i.e., one needs to remove the entire system for the system to collapse). Therefore, a different α with fixed $\langle k_{total}\rangle$ can completely change the system's resilience to localized attacks. Remarkably, networks with the same $\langle k_{total}\rangle$ but a larger interconnectivity ratio α can be more vulnerable to localized attacks, compared to networks with small α, where the communities are not well connected but are more self-sufficient. This result suggests that when a network is designed (as usually happens in infrastructures, for example) with fixed $\langle k_{total}\rangle$, it is of interest to minimize α to make the system more robust to localized attacks.

6 Conclusions

Complex networks appear and affect our lives in diverse and unique ways, from biochemical activities in our protein–protein interactions network or our brain to infrastructures or transportation networks we use for traveling. Percolation theory is found to be useful for studying the resilience of such networks under random failure or localized attack. The network's resilience is significantly affected by its spatial constraints, which are expressed in different properties of the network such as the distribution of edge lengths in space, spatial modular structure, or the length of the dependency links. Understanding these spatial properties is crucial for preventive actions and mitigation against

(a) (b)

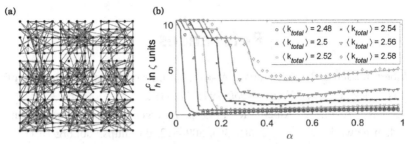

Figure 13 Localized attack on interdependent heterogeneous modular model (a) Two heterogeneous modular networks with structure illustrated in Fig. 3c are fully interdependent. Nodes with the same geometric position in the two networks are interdependent, which enables us to view the networks from the top like a single network with two types of connectivity links in both networks. The green links belong to network A, and the blue to network B. **(b)** The critical radius, r_h^c, of this heterogeneous modular model. For every value of $\langle k_{total}\rangle$ a critical value of α exists, above which r_h^c drops abruptly. For $\alpha > \alpha_c$ we have a metastable regime, where a finite-size localized attack larger than r_h^c (which is independent of the system size) causes cascading failures, leading to a system collapse. In this regime the critical radius r_h^c depends weakly on the interconnectivity parameter α. In marked contrast, for $\alpha < \alpha_c$, the critical attack $r_h^c \sim 0.5L$ (i.e., one needs to remove the entire system for the system to collapse). **Source:** Vaknin et al. 2020.

different types of failures. For example, it is critical to know if the characteristic edge length of infrastructure network of networks is greater than its critical characteristic length and random failures or localized attacks can cause catastrophic cascading failures and abrupt collapse, or less than the critical characteristic length, which is characterized by a continuous transition. Another example is the assignment of links within and between communities in interdependent spatial modular networks, which significantly affects robustness to localized attacks. All these spatial properties should be taken into account when a network is designed or when attempting to improve network resilience, in case it already exists. If the world's awareness of the strict relation between network resilience and its spatial properties rises, we hope to see fewer catastrophic events of network collapse, which will improve the quality of our life.

In this Element, we briefly introduced recent developments in the study of percolation in spatial networks. However, the analytical results are few, and much effort is still needed to improve our understanding in this area. Advanced spatial network models whose analytical solvable are still missing, and their resilience properties may be the next important breakthrough in percolation theory.

References

Albert, R., Jeong, H., & Barabási, A.-L. (2000). Error and attack tolerance of complex networks. *Nature, 406*(6794), 378–382.

Barabási, A.-L., & Albert, R. (1999). Emergence of scaling in random networks. *Science, 286*(5439), 509–512. doi: https://doi.org/10.1126/science.286.5439.509.

Barthélemy, M. (2011). Spatial networks. *Physics Reports, 499*(1–3), 1–101.

Bashan, A., et al. (2013). The extreme vulnerability of interdependent spatially embedded networks. *Nature Physics, 9*(10), 667–672.

Bell, M. G., & Iida, Y. (1997). *Transportation network analysis.* Wiley Online Library.

Ben-Avraham, D., et al. (2003). Geographical embedding of scale-free networks. *PhysicaA: Statistical Mechanics and Its Applications, 330*(1–2), 107–116.

Berezin, Y., et al. (2015). Localized attacks on spatially embedded networks with dependencies. *Scientific Reports, 5*(1), 1–5.

Bianconi, G., Pin, P., & Marsili, M. (2009). Assessing the relevance of node features for network structure. *Proceedings of the National Academy of Sciences, 106*(28), 11433–11438.

Boguñá, M., et al. (2021). Network geometry. *Nature Reviews Physics*, Nature Publishing Group *3* (2), 114–135.

Bollobás, B. (1985). *Random Graphs.* London Mathematical Society Monographs, Academic Press, London.

Bonamassa, I., et al. (2019). Critical stretching of mean-field regimes in spatial networks. *Physical Review Letters, 123*(8), 088301.

Buldyrev, S. V., et al. (2010). Catastrophic cascade of failures in interdependent networks. *Nature, 464*(7291), 1025–1028.

Bullmore, E., & Sporns, O. (2012). The economy of brain network organization. *Nature Reviews Neuroscience, 13*(5), 336–349.

Bunde, A., & Havlin, S. (1991). Fractals and disordered systems. Springer, New York.

Cohen, R., Havlin, S., & Ben-Avraham, D. (2003). Efficient immunization strategies for computer networks and populations. *Physical Review Letters, 91*(24), 247901.

Cohen, R., et al. (2000). Resilience of the internet to random breakdowns. *Physical Review Letters, 85*(21), 4626.

Cohen, R., et al. (2001). Breakdown of the internet under intentional attack. *Physical Review Letters, 86*(16), 3682.

Colizza, V., Barrat, A., Barthélemy, M., & Vespignani, A. (2006). The role of the airline transportation network in the prediction and predictability of global epidemics. *Proceedings of the National Academy of Sciences, 103*(7), 2015–2020.

Danziger, M. M., et al. (2013). Interdependent spatially embedded networks: dynamics at percolation threshold. In *2013 International Conference on Signal-Image Technology & Internet-Based Systems* (pp. 619–625). doi: https://doi.org/10.1109/SITIS.2013.101

Danziger, M. M., et al. (2016). The effect of spatiality on multiplex networks. *EPL (Europhysics Letters), 115*(3), 36002.

Danziger, M. M., et al. (2020). Faster calculation of the percolation correlation length on spatial networks. *Physical Review E, 101*(1), 013306.

Daqing, L., et al. (2011). Dimension of spatially embedded networks. *Nature Physics, 7*(6), 481–484.

Den Nijs, M. (1979). A relation between the temperature exponents of the eight-vertex and q-state Potts model. *Journal of Physics A: Mathematical and General, 12*(10), 1857.

Dong, G., et al. (2018). Resilience of networks with community structure behaves as if under an external field. *Proceedings of the National Academy of Sciences, 115*(27), 6911–6915.

Donges, J. F., et al. (2009). Complex networks in climate dynamics. *The European Physical Journal Special Topics, 174*(1), 157–179.

Dosenbach, N. U., et al. (2007). Distinct brain networks for adaptive and stable task control in humans. *Proceedings of the National Academy of Sciences, 104*(26), 11073–11078.

Emmerich, T., et al. (2013). Complex networks embedded in space: dimension and scaling relations between mass, topological distance, and Euclidean distance. *Physical Review E, 87*(3), 032802.

Ercsey-Ravasz, M., et al. (2013). A predictive network model of cerebral cortical connectivity based on a distance rule. *Neuron, 80*(1), 184–197.

Erdős, P., & Rényi, A. (1959). On random graphs I. *Publicationes Mathematicae Debrecen, 6*, 290.

Erdős, P., & Rényi, A. (1960). On the evolution of random graphs. *Publications of the Mathematical Institute of the Hungarian Academy of Sciences, 5*(1), 17–60.

Faloutsos, M., Faloutsos, P., & Faloutsos, C. (1999). On power-law relationships of the internet topology. *ACM SIGCOMM Computer Communication Review, 29*(4), 251–262.

Fan, J., et al. (2017). Network analysis reveals strongly localized impacts of El Niño. *Proceedings of the National Academy of Sciences, 114*(29), 7543–7548.

Fan, J., et al. (2018). Structural resilience of spatial networks with inter-links behaving as an external field. *New Journal of Physics, 20*(9), 093003.

Gallos, L. K., et al. (2012). A small world of weak ties provides optimal global integration of self-similar modules in functional brain networks. *Proceedings of the National Academy of Sciences, 109*(8), 2825–2830.

Gao, J., Li, D., & Havlin, S. (2014). From a single network to a network of networks. *National Science Review, 1*(3), 346–356.

Gao, J., et al. (2011). Robustness of a network of networks. *Physical Review Letters, 107*(19), 195701.

Gao, J., et al. (2012). Networks formed from interdependent networks. *Nature Physics, 8*(1), 40.

Gastner, M. T., & Newman, M. E. (2006). The spatial structure of networks. *The European Physical Journal B-Condensed Matter and Complex Systems, 49*(2), 247–252.

Gibson, T. E., et al. (2016). On the origins and control of community types in the human microbiome. *PLoS Computational Biology, 12*(2), e1004688.

Goldenberg, J., & Levy, M. (2009). Distance is not dead: social interaction and geographical distance in the internet era. *arXiv preprint arXiv:0906.3202*.

Gross, B., & Havlin, S. (2020). Epidemic spreading and control strategies in spatial modular network. *Applied Network Science, 5*(1), 1–14.

Gross, B., et al. (2017). Multi-universality and localized attacks in spatially embedded networks. In *Proceedings of the Asia-Pacific Econophysics Conference 2016 – Big Data Analysis and Modeling toward Super Smart Society – (APEC-SSS2016) JPS Conference Proceedings*, **16**, 011002. doi: https://doi.org/10.7566/JPSCP.16.01100

Gross, B., et al. (2020a). Interconnections between networks acting like an external field in a first-order percolation transition. *Physical Review E, 101*(2), 022316.

Gross, B., et al. (2020b). Two transitions in spatial modular networks. *New Journal of Physics, 22*(5), 053002.

Gross, B., et al. (2021). Interdependent transport via percolation backbones in spatial networks. *Physica A: Statistical Mechanics and Its Applications, 567*(9), 125644.

Guimera, R., Mossa, S., Turtschi, A., & Amaral, L. N. (2005). The worldwide air transportation network: anomalous centrality, community structure, and cities' global roles. *Proceedings of the National Academy of Sciences, 102*(22), 7794–7799.

Hajdu, L., Bóta, A., Krész, M., Khani, A., & Gardner, L. M. (2019). Discovering the hidden community structure of public transportation networks. *Networks and Spatial Economics, 20*(1), 209–231.

Halu, A., Mukherjee, S., & Bianconi, G. (2014). Emergence of overlap in ensembles of spatial multiplexes and statistical mechanics of spatial interacting network ensembles. *Physical Review E, 89*(1), 012806.

Hamedmoghadam, H., etal. (2021). Percolation of heterogeneous flows uncovers the bottlenecks of infrastructure networks. *Nature Communications, 12*(1), 1–10.

Havlin, S., & Nossal, R. (1984). Topological properties of percolation clusters. *Journal ofPhysics A: Mathematical and General, 17*(8), L427.

Horvát, S., et al. (2016). Spatial embedding and wiring cost constrain the functional layout of the cortical network of rodents and primates. *PLoS biology, 14*(7), e1002512.

Hu, Y., et al. (2011). Possible origin of efficient navigation in small worlds. *Physical Review Letters, 106*(10), 108701.

Huang, W., et al. (2014). Navigation in spatial networks: a survey. *Physica A: Statistical Mechanics and Its Applications, 393*, 132–154.

Jeong, H., et al. (2000). The large-scale organization of metabolic networks. *Nature, 407*(6804), 651–654.

Kirkpatrick, S. (1973). Percolation and conduction. *Reviews of Modern Physics, 45*(4), 574.

Kleinberg, J. M. (2000). Navigation in a small world. *Nature, 406*(6798), 845.

Kovács, I. A., et al. (2019). Network-based prediction of protein interactions. *Nature Communications, 10*(1), 1–8.

Lambiotte, R., et al. (2008). Geographical dispersal of mobile communication networks. *Physica A: Statistical Mechanics and Its Applications, 387*(21), 5317–5325.

Latora, V., & Marchiori, M. (2005). Vulnerability and protection of infrastructure networks. *Physical Review E, 71*(1), 015103.

Li, D., et al. (2011). Percolation of spatially constraint networks. *EPL (Europhysics Letters), 93*(6), 68004.

Li, D., et al. (2015). Percolation transition in dynamical traffic network with evolving critical bottlenecks. *Proceedings of the National Academy of Sciences, 112*(3), 669–672. Retrieved from www.pnas.org/content/112/3/669.abstract doi: https://doi.org/10.1073/pnas.1419185112.

Li, Z., et al. (2017). The OncoPPi network of cancer-focused protein-protein interactions to inform biological insights and therapeutic strategies. *Nature Communications, 8*(1), 1–14.

Liben-Nowell, D., Novak, J., Kumar, R., Raghavan, P., & Tomkins, A. (2005). Geographic routing in social networks. *Proceedings of the National Academy of Sciences, 102*(33), 11623–11628.

Liu, Y., et al. (2021). Efficient network immunization under limited knowledge. *National Science Review, 8*(1), nwaa229.

Ludescher, J., et al. (2014). Very early warning of next El Niño. *Proceedings of the National Academy of Sciences, 111*(6), 2064–2066.

Markov, N. T., et al. (2014). A weighted and directed interareal connectivity matrix for macaque cerebral cortex. *Cerebral Cortex, 24*(1), 17–36.

Menck, P. J., Heitzig, J., Kurths, J., & Schellnhuber, H. J. (2014). How dead ends undermine power grid stability. *Nature Communications, 5*(1), 1–8.

Milgram, S. (1967). The small world problem. *Psychology Today, 2*(1), 60–67.

Milo, R., et al. (2002). Network motifs: simple building blocks of complex networks. *Science, 298*(5594), 824–827.

Moretti, P., & Muñoz, M. A. (2013). Griffiths phases and the stretching of criticality in brain networks. *Nature Communications, 4,* 2521.

Newman, M. E. (2003). The structure and function of complex networks. *SIAM Review, 45*(2), 167–256.

Newman, M. E. J., Strogatz, S. H., & Watts, D. J. (2001, July). Random graphs with arbitrary degree distributions and their applications. *Phys. Rev. E, 64,* 026118.

Nienhuis, B. (1982). Analytical calculation of two leading exponents of the dilute Potts model. *Journal of Physics A: Mathematical and General, 15*(1), 199.

Paine, R. T. (1966). Food web complexity and species diversity. *The American Naturalist, 100*(910), 65–75.

Parshani, R., etal. (2010). Interdependent networks: reducing the coupling strength leads to a change from a first to second order percolation transition. *Physical Review Letters, 105*(4), 048701.

Pastor-Satorras, R., et al. (2015). Epidemic processes in complex networks. *Reviews of Modern Physics, 87*(3), 925.

Pocock, M. J., et al. (2012). The robustness and restoration of a network of ecological networks. *Science, 555*(6071), 973–977.

Polis, G. A., & Strong, D. R. (1996). Food web complexity and community dynamics. *The American Naturalist, 147*(5), 813–846.

Reis, S. D., et al. (2014). Avoiding catastrophic failure in correlated networks of networks. *Nature Physics, 10*(10), 762–767.

Reynolds, P., Stanley, H., & Klein, W. (1977). Ghost fields, pair connectedness, and scaling: exact results in one-dimensional percolation. *Journal of Physics A: Mathematical and General, 10*(11), L203.

Rozenfeld, A. F., et al. (2002). Scale-free networks on lattices. *Physical Review Letters, 89*(21), 218701.

Schmeltzer, C., Soriano, J., Sokolov, I. M., & Rudiger, S. (2014). Percolation of spatially constrained Erdős-Rényi networks with degree correlations. *Physical Review E, 89*(1), 012116.

Shai, S., et al. (2015). Critical tipping point distinguishing two types of transitions in modular network structures. *Physical Review E, 92*(6), 062805.

Shao, S., et al. (2015). Percolation of localized attack on complex networks. *New Journal of Physics, 17*(2), 023049.

Shekhtman, L. M., et al. (2014). Robustness of a network formed of spatially embedded networks. *Physical Review E, 90*(1), 012809.

Smillie, C. S., et al. (2011). Ecology drives a global network of gene exchange connecting the human microbiome. *Nature, 480*(7376), 241–244.

Sporns, O. (2010). *Networks of the brain.* Massachusetts Institute of Technology Press.

Stanley, H. (1971). *Introduction to phase transitions and critical phenomena.* Oxford University Press.

Stauffer, D., & Aharony, A. (2018). *Introduction to percolation theory.* CRC Press.

Stauffer, D., & Sornette, D. (1999). Self-organized percolation model for stock market fluctuations. *Physica A: Statistical Mechanics and Its Applications, 271*(3–4), 496–506.

Stippinger, M., & Kertész, J. (2014). Enhancing resilience of interdependent networks by healing. *Physica A: Statistical Mechanics and Its Applications, 416,* 481–487.

Stork, D., & Richards, W. D. (1992). Nonrespondents in communication network studies: problems and possibilities. *Group & Organization Management, 17*(2), 193–209.

Suki, B., Bates, J. H., & Frey, U. (2011). Complexity and emergent phenomena. *Comprehensive Physiology, 1*(2), 995–1029.

Sykes, M. F., & Essam, J. W. (1964). Exact critical percolation probabilities for site and bond problems in two dimensions. *Journal of Mathematical Physics, 5*(8), 1117–1127.

Vaknin, D., et al. (2017). Spreading of localized attacks in spatial multiplex networks. *New Journal of Physics, 19*(7), 073037.

Vaknin, D., et al. (2020). Spreading of localized attacks on spatial multiplex networks with a community structure. *Physical Review Research, 2*(4), 043005.

Viswanathan, G. M., et al. (1999). Optimizing the success of random searches. *Nature, 401*(6756), 911–914.

Wang, W., et al. (2017). Unification of theoretical approaches for epidemic spreading on complex networks. *Reports on Progress in Physics, 80*(3), 036603.

Waxman, B. M. (1988). Routing of multipoint connections. *IEEE Journal on Selected Areas in Communications, 6*(9), 1617–1622.

Wei, L., et al. (2012). Cascading failures in interdependent lattice networks: the critical role of the length of dependency links. *Physical Review Letters, 108*(22), 228702.

Wei, L., et al. (2014). Ranking the economic importance of countries and industries. *Journal of Network Theory in Finance, 3*(3), 1–17.

Yang, Y., Nishikawa, T., & Motter, A. E. (2017). Small vulnerable sets determine large network cascades in power grids. *Science*, American Association for the Advancement of Science *358*(6365), eaan3184.

Yuan, X., Shao, S., Stanley, H. E., & Havlin, S. (2015). How breadth of degree distribution influences network robustness: comparing localized and random attacks. *Physical Review E, 92*(3), 032122.

Zhou, D., et al. (2014). Simultaneous first- and second-order percolation transitions in interdependent networks. *Physical Review E, 90*(1), 012803.

Acknowledgments

We thank the Israel Science Foundation, the Binational Israel-China Science Foundation Grant No. 3132/19, ONR, the BIU Center for Research in Applied Cryptography and Cyber Security, NSF-BSF Grant No. 2019740, the EU H2020 project RISE (Project No. 821115), the EU H2020 DIT4TRAM, and DTRA Grant No. HDTRA-1-19-1-0016 for financial support.

Bnaya dedicates this book to his beloved wife Tsofia for her calm patience and endless support.

Cambridge Elements ≡

The Structure and Dynamics of Complex Networks

Guido Caldarelli

Ca' Foscari University of Venice

Guido Caldarelli is Full Professor of Theoretical Physics at Ca' Foscari University of Venice. Guido Caldarelli received his Ph.D. from SISSA, after which he held postdoctoral positions in the Department of Physics and School of Biology, University of Manchester, and the Theory of Condensed Matter Group, University of Cambridge. He also spent some time at the University of Fribourg in Switzerland, at École Normale Supérieure in Paris, and at the University of Barcelona. His main scientific activity (interest?) is the study of networks, mostly analysis and modelling, with applications from financial networks to social systems as in the case of disinformation. He is the author of more than 200 journal publications on the subject, and three books, and is the current President of the Complex Systems Society (2018 to 2021).

About the Series

This cutting-edge new series provides authoritative and detailed coverage of the underlying theory of complex networks, specifically their structure and dynamical properties. Each Element within the series will focus upon one of three primary topics: static networks, dynamical networks, and numerical/computing network resources.

Cambridge Elements ⊒

The Structure and Dynamics of Complex Networks

Printed in the United States
by Baker & Taylor Publisher Services